FIGHTERS

Created and produced by Firecrest Books Ltd
in association with John Francis/Bernard Thornton Artists

Copyright © 2000 Firecrest Books Ltd
and copyright © 2000 John Francis/Bernard Thornton Artists

Published by Tangerine Press™, an imprint of Scholastic Inc.
555 Broadway, New York, NY 10012

Tangerine Press™ and associated logo and design are trademarks of Scholastic Inc.

ISBN 0-439-20657-X

Printed and bound in Belgium
First printing September 2000

FIGHTERS

Bernard Stonehouse

Illustrated by
John Francis

TANGERINE PRESS™ and associated logo
and design are trademarks of Scholastic Inc.

For Bernard Thornton

Art and Editorial Direction by
Peter Sackett

Designed by
Paul Richards, Designers & Partners

Edited by
Norman Barrett

Color separation by
Sang Choy International Pte. Ltd.
Singapore

Printed and bound by
Casterman, Belgium

— Contents —

Introduction 6

Stag beetles 8

Bighorn sheep 10

African elephants 12

Siamese fighting fish 14

Giraffes 16

Brown hares 18

Jungle fowl 20

Moose 22

Olive baboons 24

Poison dart frogs 26

Warthogs 28

Cassowaries 30

African lions 32

European robins 34

Sticklebacks 36

Red kangaroos 38

Ruffs 40

Timber wolves 42

Chimpanzees 44

Burchell's zebras 46

Index 48

— Introduction FIGHT AND RUN AWAY

We like to think that "wild" animals are truly wild, fighting constantly among themselves with teeth and claws, growls and snarling. Visitors to game parks are often disappointed to find "fierce" lions staring contentedly at the horizon, gorillas snoring together in comfortable heaps, even wolves sniffing each other like friendly dogs. Like most other animals, they are smart enough to fight as little as possible.

If animals fought constantly with teeth, claws, and all their strength, there would be very few left. Nearly all fight at some time in their lives, and fights can be fierce and sometimes end in death. However, fighting takes a lot of energy and is liable to damage both parties. If the loser dies today, and the winner dies tomorrow from infected bites or wounds, then both have lost. It is far better to know how to fight, to show that you know, and yet to avoid fighting.

So most fights between animals, such as those in this book, follow rituals — like boxing matches. There is much noise and show of strength, but not always a lot of bloodshed. The winner stands its ground, the loser slinks away, having learned — if it is capable of learning — that fighting that particular opponent does not pay. Both live to fight another day.

Angry elephant – but is he
ready to fight?

Stag beetles ANTLERLIKE JAWS

Here is a wrestling match between two beetles, each about 2 inches (5 cm) long. They are called stag beetles because the jaws look vaguely like antlers — the big, bony outgrowths that stags (male deer) carry on their heads in spring and summer. Males are larger than females, with bigger and stronger jaws.

Stag beetles come in all colors, from black to bright green, red, or yellow. You find them scuttling under fallen trees and among rotting stumps, where they feed, mate, and lay their eggs. Despite their weight and clumsiness, stag beetles fly well. Under those shiny back plates are wings, which they unfold and spread on warm evenings, buzzing heavily through the forest and dropping to the ground in search of a place to live.

Home for a male stag beetle is a small patch of tree trunk or woodland floor, where it feeds and tries to keep other males away. When two meet, they rear up and circle around each other, as though assessing their chances, and try to grasp each other with their jaws. The fight ends when one lifts the other into the air, and drops it on its back, to lie helplessly for a few moments with legs waving. Then both go their separate ways.

Bighorn sheep BATTERING RAMS

Bighorns are wild sheep that live high in the western mountains of North America. Ewes (females) have small, spiky horns that point backward from the top of the skull. Rams (males) have much bigger horns that curve downward on either side of the head. The horns grow a few inches a year, so a fully grown ram of 10 years old has a truly massive set, hard and bony, reaching down below his eyes.

Horns show the age and sex of the "wearer." In autumn, at the start of the breeding season, both females and young males keep out of the way of the males with big horns. The big males meet in pairs or groups and charge each other head-on, crashing their horns together with a thump that echoes among the peaks. The females charge each other in the same way, but less resoundingly. This ritualized fighting helps to keep the bighorn population well spaced in their habitat. If we bashed each other around like that, we would probably end up with headaches and cracked skulls. Bighorn sheep do it all day, and none seem the worse for it.

African elephants TOO BIG TO FIGHT

African elephants live in groups of a dozen or more, each group including several females with calves and one or two young males. Older males wander on their own, joining the herds only for a few days at a time, often when one of the females is ready to breed.

Elephants rarely fight. They are too big and heavy to throw their weight around, and quickly overheat if they try. If two males meet as rivals in the presence of a female, they settle their differences by edging each other out, and eventually by a pushing match. They stand head-on, with tusks raised, trunks swinging, and ears extended. Scraping the ground with massive forefeet, sometimes trumpeting and squealing, they approach each other as though ready for the battle to end all battles. Then they lock trunks and push hard. Tusks may interlock and occasionally crack, but the intention is a show of strength rather than a damaging fight.

Who wins? After a time, one seems to recognize that the other is stronger, so he backs away and disengages in a "gentlemanly" sort of way. He may still hang around with the herd, but he keeps his place as a subordinate member. Next time he'll be just a little bigger and stronger, and perhaps more successful.

Siamese fighting fish BRED FOR COMBAT

Many kinds of fish fight each other for space or a mate, but few do it as elegantly or persistently as these colorful freshwater fish of Southeast Asia. In the wild, they are relatively drab little fish 2 inches (5 cm) long, found in streams and ponds. However, early in the breeding season, the males become more colorful. They each blow a small canopy of bubbles, making an upside-down nest at the surface of the water. The male invites females into his nest to lay their eggs, and excludes other males by postures, threatening, and, if necessary, fighting.

In open water where space is plentiful, there are more threats than fights. Confined to a jar or tank, rivals cannot get away from each other and may fight continuously, biting fins and scraping each other's flesh until one of them dies of exhaustion. In Asia, they have been bred for their fighting qualities.

The fish shown here have been specially bred to be colorful and show fighting spirit. Some will fight rivals for hours on end, winning bets and earning prize money for their owners. If we do not want them to fight, we have to keep them out of each other's reach when the breeding season comes around.

In the wild, in the non-breeding season, the male Siamese fighting fish is a drab individual.

Giraffes LOOKING DOWN ON THE WORLD

Giraffes look down on the world from a great height. A large, fully grown male may stand 18 feet (5.5 m) tall — much taller than any other animal of the African plains. Fully grown females stand slightly smaller, but are still taller than any rival species. The long neck provides half the height, the long forelegs and barrellike body the rest. Few other animals are half as graceful or elegant.

Standing the height of a telephone pole has both advantages and problems. A giraffe can see long distances across the plains and, with its long tongue, grasp leaves and shoots of trees beyond the reach of other ground-dwelling animals. Even half-grown, it has very few predators to worry about. Long legs can carry it out of danger quickly, and enable it to kick hard when there is a need. The problems? It's the wrong shape for pushing and butting.

So can it fight rival giraffes? Yes, and with all the grace you would expect. Its head is hard, with bony crests, and its neck is packed solid with muscle. Rival giraffes stand shoulder to shoulder, swinging their heads and necks around each other like clubs. Sooner or later one decides that it has had enough, and the battle is over.

Brown hares MARCH MADNESS

Like large, long-legged rabbits, brown hares graze and browse in grasslands, meadows, and along the forest edges. They are widespread across Europe and Asia, with close kin in Africa and North America. Unlike rabbits, they live alone or in small, widely spaced groups of individuals. They never burrow, but rely on camouflage to hide themselves. Litters of three or four tawny leverets (young hares), nesting among leaf litter and long grass, match their background almost perfectly. Adults rest during the day, feeding mainly in the evenings and early mornings. Approached by a predator, they lie still until the last moment, then leap up and race off on their long, powerful legs.

Male hares, called "jacks ," live quietly during the winter when food is scarce. In early spring, their behavior changes remarkably. They gather in groups of up to a dozen, often on a hilltop or prominent place, leaping and dancing as though demented. Facing each other, they box with their forearms, then leap over each other, striking hard with their hind legs. There is seldom a recognizable winner, but somehow these antics seem to determine which of them will mate, or mate most often, with the admiring females nearby.

A young hare, or leveret, lies quietly and well camouflaged in an open nest.

— Jungle fowl DEADLY SPURS, FLYING FEATHERS

They look like a couple of barnyard roosters — not surprisingly, because these red jungle fowl from the forests of Southeast Asia are the original stock from which domestic poultry were derived. The red comb, fleshy cheek lappets (flaps), and yellow bill are typical, as are the red-and-yellow hackles that protect the neck, and the black body plumage shot with iridescent green.

Jungle fowl and domestic fowl behave in many similar ways. The business of the rooster is to round up small groups of hens, mate with them from time to time, and defend them against the attentions of other roosters. Crowing advertises their presence. Meeting other males makes them aggressive, and when necessary they fight.

Fighting males square off with each other, leaping into the air and striking with inch-long (2.5 cm) bony spurs on the inside of their legs. In the wild or in the farmyard, fights seldom last more than a few minutes. Feathers may fly, but after two or three rounds, one retreats and the other crows to announce his victory, with little harm done to either.

Moose SPIKED ANTLERS

Largest of all the deer, a fully grown bull moose stands over 7 feet (2 m) tall at the shoulder, and weighs over 1,500 pounds (680 kg). Moose are found most often in wet woodlands. Their favorite food is the soft aquatic vegetation that grows in lakes and ponds. Typically they stand shoulder-high in the water, dipping their huge heads to browse. In winter, when the ponds are ice-covered, they dig in the snow for young wood and bark.

Cow moose live in small groups with their growing young, while the bulls wander off on their own. They come together in September and October, when the bulls are wearing their new sets of spiked antlers and the cows are ready for mating. Cows roar to attract the bulls. If two or more bulls respond, bashing their way through the forest toward the same cow, there may well be trouble. Heads are lowered, antlers interlocked, and the two males push backward and forward like overcharged steam engines on the same piece of track. Which one wins? The one that pushes harder and longer.

The moose lives in wild areas of North America and Asia and in northern Europe, where it is known as an elk.

Olive baboons A SHOW OF TEETH

Here is a good 80 pounds (36 kg) of solid male baboon, with a face full of sharp ivory teeth, looking like he's ready to tear pieces from a rival male. What is this all about?

Olive baboons live in mountains and scrubland of Central Africa, in mixed troops of males, females, and young. Males normally live in harmony, tolerating each other's company and often hunting together. When packs of dogs or hyenas threaten, they unite to attack the predators and lead the troop to safety. This fight is not half as serious as it looks. The younger male, with his back to us, is trying to join the troop. The older one sees him as a possible rival. He has shown his teeth and thumped the ground with his paws. Now he is charging. If the intruder runs off, the attacker wins. If the intruder stands his ground, there will be a scuffle, some angry squealing, and possibly a painful bite or two, before one or the other turns away. Then all will be peaceful again.

Baboons live in troops of 60 to 80 individuals, including males, females, and young. They feed together and groom each other's fur.

— Poison dart frogs WRESTLING ON A LEAF

What do frogs find to fight about? Here are a couple locked in combat, like miniature Sumo wrestlers, and trying to throw each other off a narrow leaf.

These are poison dart frogs from Central America. Their brilliant colors are a warning to snakes and other predators not to touch them, for their skin contains poisons. Human hunters have for long known how to collect the poisons and transfer them to the points of arrows or darts, which they use to kill other animals. A bird or small mammal grazed by the tip of a poisoned dart dies quickly.

What looks like a fight is more like a courtship dance. At breeding time, pregnant females and males gather in groups in damp corners of the forest and leap, jump, and wrestle together. This seems to stimulate the females to lay their eggs on the ground, or on the surface of a leaf, and the males to fertilize them. Then each male attaches a dozen or more eggs to his own back, and carries them around until they hatch into tadpoles.

— Warthogs AFRICAN PIG

A warthog is an African pig with a tawny mane, a broad, flattened snout, a collection of lumps, bumps, scales, and warts on its face, two pairs of tusks, and an uncertain temper. Warthogs live in Africa, usually close to rivers or lakes. Sows and their young form small family groups of a dozen or more. Hogs (males) live in small bachelor groups or alone, joining the females briefly for mating, then wandering off on their own. Warthogs browse on grass, dig for roots, and usually live at peace with one another and the rest of the world.

Both males and females carry those long, curly tusks and use them for rooting and digging up the ground. Both can defend themselves stoutly against lions, hyenas, and other predators, using their weight and their tusks, which they slash from side to side like bowie knives. On rare occasions when males fight each other, the tusks are prominent but seldom used. Slashing would be far too dangerous. Rival males charge each other like tanks, grunting and squealing, each trying to knock the other sideways and bowl him over.

Cassowaries HARD HEADS AND SHARP TOES

Tall as a man, weighing about 150 pounds (70 kg) or more, cassowaries live in the dense rain forests of northern Australia, New Guinea, and neighboring islands. It is a tough environment. The coarse, wiry feathers covering the back and flanks of these flightless birds protect them from thorns and sharp-edged leaves. The skull is capped with a hard, bony ridge like a helmet — useful protection when they push headfirst through the undergrowth. Their legs are long and bumpy, their feet flat and three-toed. Too heavy to fly, they use their small wings for balance when they run. They feed on fruit, leaves, and insects gathered from shrubs and the forest floor.

Cassowaries seldom fight, but the males, which incubate the eggs and care for the chicks, have a reputation for dealing hard with both predators and intruders of their own species that come too close to the family. Their weapon is the big inner toe, with a razor-sharp nail, on either foot. A fighting cassowary leaps upward like a jungle fowl (see page 20), slashing down with one or both feet. It is an argument that others quickly learn to respect.

Cassowaries live in small groups deep in the forest. The male alone incubates and cares for the chicks.

— African lions LAZY BUT DANGEROUS

The lions that tourists see in game parks and reserves are sometimes a disappointment. Storybooks tell us that these are the fiercest of beasts — powerful, bold, and ready to take on all challengers. Yet there they lie, dozing in the shade, too lazy even to flick the flies from their noses.

Lions in game parks are often surrounded by antelopes and other prey. A pride of nine or a dozen, including an old male, two or three young males, and several females with their cubs, can take all the food they need in a few hours of evening hunting, leaving plenty of time for being lazy. They hunt cooperatively, sharing the food between them. Occasionally the males spar together like overgrown kittens, but there is very little serious fighting.

During spells of drought, when game becomes scarce and wary, the prides may split up and become smaller. Sparring between the males becomes more serious. With claws and teeth fully bared, the biggest lion, having the shaggiest mane and strongest sense of belonging, may drive the younger males out. Or he may wander off on his own, to join the same group or another when the difficult times are over.

Lion (far left), lioness, and cub rest in the shade of a game park.

European robins BORDER DISPUTES

Robin redbreasts of the kind shown here are restricted mainly to Europe, but are well known all over the world. "Sweet little bird in russet coat," sang the poet, and Christmas cards the world over celebrate robins as symbols of peace and goodwill.

Part of their charm is tameness, particularly in winter when food is scarce. Long ago, robins learned the trick of pecking over ground that larger animals — such as hogs — have recently disturbed. The modern equivalent is the gardener with his spade or trowel. Turn the soil over, and your friendly neighborhood robin sitting sharp-eyed on the fence darts almost underfoot to pick up grubs and insects.

Though tolerant of humans, robins in winter have no time for each other. Males and females, which look very much alike, take up separate territories in autumn and defend them fiercely against each other. Along the borders they threaten rivals with their upward-pointing bills, showing off their orange-red breasts. If this is not enough, they fly at each other with bill and claws. Almost inevitably, one or the other quickly backs off. In spring, the mood changes. Males and females in neighboring territories often become mates, raising two or three broods in summer harmony.

The sharp-eyed robin waits and watches for its next meal.

Sticklebacks SPINY FISH THAT DEFEND THEIR NESTS

Two-spined sticklebacks of North America and three-spined sticklebacks of Europe (shown here) are similar and closely related fishes, common in clear, shallow freshwater streams. About 3 inches (7.5 cm) long, they are insignificant little fishes during autumn and winter, barely visible from above against a background of gravel or mud. In spring their appearance changes. Males particularly shift into clear metallic greens and blues along the back and flanks, and the throat and belly become spectacular orange-red. Females also grow brighter, and lose their streamlined shape as eggs accumulate, develop, and ripen inside their bodies.

Their behavior changes, too. Males drop to the streambed and space themselves apart in small territories. Each gathers sprigs of water plants, pulling them into a nest and gluing them together with a liquid secretion from the kidneys. The owner defends the area immediately around the nest. Any orange-throated male that he sees is fiercely fought off with gaping jaws and lashing tail. That is what is happening here.

A pregnant female by contrast is warmly welcomed. With a zigzag dance, the male leads her toward the nest, and nudges her gently to lay her eggs within it. Then he chases her away, and guards and tends the eggs alone until they hatch.

Red kangaroos KICK-BOXERS OF THE OUTBACK

These male red kangaroos have found something to quarrel over, and are slugging it out with a kind of kick-boxing. They live in Australia, grazing in small groups, or "mobs," on dry, dusty grasslands of the outback. Large males like these, with reddish-brown woolly fur, stand up to 6 feet (1.8 m) tall and weigh as much as a heavy man. Females, blue-gray rather than red, are slightly smaller. In either sex, much of the weight is packed into the massive muscles of the hind legs and tail. The chest and forelimbs are relatively small.

A mob usually includes a dozen or more kangaroos of both sexes, young and old. When grazing quietly, they swing forward on both forelimbs and hind legs. In a hurry, they stand upright and bound through the air on their hind legs, taking leaps of 25 feet (7.6 m) or more, with the tail acting as both a counterbalance and a spring.

Most of the time they graze peacefully together. What do they fight over? Youngsters spar with each other between bouts of feeding, squaring off and boxing with their forelimbs. Mature males sometimes pitch into each other more seriously, perhaps over females, kicking out with powerful hind feet and sharp toenails.

The tail serves as a counterbalance as the kangaroo bounds along.

Ruffs COLLARS, CORONETS, AND CRAZY DANCING

Ruffs are a species of waders, or shorebirds. The males are called ruffs and the females are called reeves. In the breeding season, ruffs, about 1 foot (30 cm) long, develop a bright red face, a collar or ruff of feathers around the neck, and a coronet of plumes. Reeves are slightly smaller, without the decorations. They breed on tundra and grasslands of northern Europe and Asia, feeding on worms, crustaceans, insects, and seeds. They winter in separate flocks in warmer climates farther south.

They begin returning to the breeding grounds from early April. The ruffs, in full breeding plumage, gather on grassy mounds, tramping out small territories about 3 feet (1 m) apart, and start to show off. With collar, wings, and tail spread, head and bill pointing along the ground, they run, spin, and flutter as though demented. When two or three claim the same mound, display turns to threat, chase, and fight. Usually the older and more experienced birds hold the central sites, fighting very little, while the younger ones display and squabble around the edges. Reeves watch closely from the sidelines, eventually mating with the ruff of their choice. Once mated, they leave the area to nest, lay, and incubate their eggs, while the ruffs continue dancing, strutting, and fighting.

Timber wolves LEARNING TO LIVE TOGETHER

Here are two young wolves fighting, half in play and half for real. The one on top is winning — threatening, with lips curled to show his formidable teeth. The other has surrendered, with face turned away in submission. They are learning important lessons from each other about living together — how to win without killing, lose without being killed, take advantage of weakness, and defer to greater strength.

Wolves are widespread across Asia, North America, and northern Europe. Humans have displaced them from many of their former haunts, and today they are restricted to places where there are few, if any, people, often where food is scarce. Gray or brown, like large, rangy dogs, they are intelligent and ruthless. As meat-eaters, they cannot be choosy, and must be ready to kill whatever crosses their path. They live singly, in pairs, or in loose packs of up to a dozen or more. When game is small and scarce, they hunt alone or in small bands. When the prey is large — moose, for example — they combine into larger bands and hunt cooperatively.

Learning to cooperate can be a painful business, in which this kind of ritual fighting plays an important role.

Chimpanzees RESPECT FOR ELDERS

We think of chimpanzees as near-humans, but at home in the African forests they are simply chimpanzees, with a dignity of their own. These two are threatening and fighting as two humans may fight, but they are doing it the chimpanzee way, for chimpanzee reasons.

They belong to a loose band of 20 to 30 males, females, and young that wander through the forest in a constant search for fruit, shoots, insects, and small game — foods that vary much from season to season. Older, experienced animals learn and remember where to find food at different times of the year, and lead their bands in the right direction. Younger members follow the older ones, deferring to them, grooming them, giving them first choice of food, and cherishing and paying them respect.

There is very little squabbling between bands or within them, but here an older, gray-faced chimpanzee is teaching a sharp lesson to a youngster that has in some way broken the social code. Did the young male try to steal food, pay too much attention to a female, threaten or sass his elder? We do not know — but the older one does. The youngster will learn, and whatever it was, he'll think twice before doing it again.

A young chimp cradles an armful of fruits — are these ill-gotten gains?

Burchell's zebras STRIPES AND FLYING HOOVES

Dream of horses with strange round ears, wearing black-and-white-striped pyjamas, and you are dreaming of zebras. These are Burchell's zebras, the kind you are most likely to see in Central and East Africa. An old stallion (male), using teeth and sharp hooves, is trying to drive a younger one away.

Like wild horses, zebras live in herds of a dozen or more. Some are bachelor herds, with only young stallions. Others are family herds, with mares, foals, and yearlings, usually led by a single older stallion. Like horses, zebras move together. When lions or other predators approach, they gather with ears pricked and all eyes watching. Then, as likely as not, they turn and gallop off to safety.

Young stallions from time to time invade the family herds, creating confusion and racing off. If some of the mares race with them, a new family herd is formed. Alert to this trick, the older stallions watch for the intruders, snort an angry warning, stamp the ground, then charge, kick, and bite. They do not always win — by nightfall this old stallion may be grazing on his own, with the herd under a new leader.

Index CREATURES AND FEATURES

A
Africa 16, 18, 24, 28, 44, 46
African elephant 12
antelope 32
antlers 8, 22
Asia 18, 22, 40, 42
Australia 30, 38

B
baboon 24
beetle 8
bighorn sheep 10
bill 34, 40
bird 20, 30, 34, 40
bite 24, 46
box/boxing 18, 38
breed/breeding season 10, 12, 14, 26, 40
Burchell's zebra 46

C
camouflage 18
cassowary 30
Central America 26
charge/charging 10, 24, 28, 46
chimpanzee 44
claws 6, 32, 34
crowing 20

D
dance/dancing 18, 26, 36, 40
deer 8, 22
dog 24, 42

E
eggs 8, 14, 26, 30, 36, 40
elephant 7, 12
elk 22
Europe 18, 22, 34, 36, 40, 42
ewe 10

F
feet 30, 38
fish 14, 36
flightless bird 30
fowl 20, 30
frog 26

G
game park 6, 32
giraffe 16
gorilla 6
grooming 44

H
hare 18
hog 28, 34
hooves 46
horns 10
horse 46
hunting 24, 32, 42
hyena 24, 28

J
jack 18
jaw 8, 36
jungle fowl 20

K
kangaroo 38
kick 16, 38, 46
kick-boxing 38

L
leveret 18
lion 6, 28, 32, 46

M
mare 46
moose 22, 42

N
neck 16
nest 14, 18, 36, 40
New Guinea 30
North America 10, 18, 22, 36, 42

P
pig 28
plumage 20
poison dart frog 26
predator 16, 18, 24, 26, 28, 30, 46
prey 32, 42
push/pushing 12, 16, 22

R
rain forest 30
ram 10
reeve 40
ritual 6, 10, 42
robin 34
ruff 40

S
sheep 10
shorebird 40
Siamese fighting fish 14
snake 26
Southeast Asia 14, 20
spurs 20
stag beetle 8
stallion 46
stickleback 36
strength, show of 6, 12

T
tadpole 26
tail 36, 38
teeth 6, 24, 32, 42, 46
territory 34, 36
threat/threatening 14, 34, 40, 42, 44

toes/toenails 30, 38
trunk 12
tundra 40
tusks 12, 28

W
wader 40
warthog 28
wings 8, 30
wolf 6, 42

Z
zebra 46